THE EASY WAY
TO
QUIT
SUGAR

Allen Carr

WITH BEV AISBETT

THE EASY WAY TO QUIT SUGAR

THE ILLUSTRATED GUIDE

ARCTURUS

ARCTURUS

This edition published in 2022 by Arcturus Publishing Limited
26/27 Bickels Yard, 151–153 Bermondsey Street,
London SE1 3HA, UK

ISBN: 978-1-78428-879-2
AD005771UK

Printed in the UK

CONTENTS

To Sam Carroll, our candle in the wind

And Florrie...

Bev Aisbett is the author and illustrator of many self-help books. Bev has previously illustrated *The Easy Way to Stop Smoking*, *The Easy Way for Women to Stop Smoking*, *The Illustrated Easy Way to Stop Smoking* and *The Illustrated Easy Way to Stop Drinking*.

ABOUT ALLEN CARR

Allen Carr was a chain-smoker for over 30 years. In 1983, after countless failed attempts to quit, he went from 100 cigarettes a day to zero without suffering withdrawal pangs, without using willpower, and without putting on weight. He realized that he had discovered what the world had been waiting for – the easy way to stop smoking – and embarked on a mission to help cure the world's smokers.

As a result of the phenomenal success of his method, he gained an international reputation as the world's leading expert on stopping smoking and his network of clinics now spans the globe. His first book, *Allen Carr's Easy Way to Stop Smoking*, has sold over 12 million copies, remains a global best-seller, and has been published in more than 40 different languages. Hundreds of thousands of smokers have successfully quit at Allen Carr's Easyway Centres where, with a success rate of over 90%, he guarantees you'll find it easy to stop or your money back.

Allen Carr's Easyway method has been successfully applied to a host of issues including weight control, alcohol and other addictions, debt, gambling, and fears. A list of the countries with an Allen Carr centre appears at the back of this book. Should you require any assistance or if you have any questions, please do not hesitate to contact your nearest centre. For more information about Allen Carr's Easyway, please visit

www.allencarr.com

THE SOUR SIDE
OF SUGAR

This book will help you to understand the truth about
BAD SUGAR and introduce you to a proven method to
cut it out of your diet. It's fine to take **GOOD SUGAR** in
the form of fruit and vegetables, but there is no secret
about the ill effects of **BAD SUGAR**, a term which in this
book covers refined sugar, processed carbohydrates, and
starchy carbohydrates.

SUGAR ADDICTION is **KILLING** people at an
unprecedented rate through a whole range of
HEALTH PROBLEMS such as:

- **OBESITY**
- **TYPE 2 DIABETES**
- **HEART DISEASE**

ENJOYING that?

> Of COURSE!
> It's nice to
> TREAT YOURSELF!

Unfortunately, I must deliver a **BITTER PILL** – that **SWEETNESS** is actually turning things very **SOUR** for you and people around the **WORLD**!

> How SO?

SUGAR ADDICTION is affecting your health on a daily basis much more than you realize. Let's face it – even on your **BEST DAYS**, don't you feel a little **UNDER PAR**?

> I have been feeling SLUGGISH!!

> And I've put on WEIGHT!

> My THINKING has been FOGGY!

9

I'm overweight and UNWELL! Is there HOPE for me?

Don't worry – you can **TURN IT AROUND** and this book is designed to help you do that. You'll soon be feeling so much **BETTER**.

But I simply can't LIVE without SUGAR! I have such a SWEET TOOTH!

When you get the full picture of just how **MUCH BETTER** life will be without "**BAD SUGAR**", we're sure you'll begin to see it **DIFFERENTLY**.

But here's some even **BETTER NEWS!** Stick with me to the end of the book and you'll realize that you're not **SACRIFICING** anything but a **HARMFUL ADDICTION!**

But sugar is just a part of EVERYDAY LIFE!

People become hooked on **SUGAR** from a very early age. Like **SMOKING**, **SUGAR** used to be seen as **HARMLESS!**

Now we are starting to see the evidence that **SUGAR** is **EQUALLY DANGEROUS**.

But surely a little sugar is just a TREAT now and then!

If it's a **TREAT** – i.e. **GOOD FOR YOU** – why do you put a **LIMIT** on it? Isn't it because you **INSTINCTIVELY** know it's doing you **HARM**?

After all, it's **BOTHERED** you enough to **PICK UP THIS BOOK**, hasn't it?

Maybe... but sugar gives me **PLEASURE**!

No doubt **YOU THINK** it does.

And it gives me a **LIFT**!

Again, that's what you've been **LED TO BELIEVE** about the **ADDICTION**.

11

Well, I wouldn't say I was **ADDICTED**! I can **LIMIT** myself using **WILLPOWER**!

WILLPOWER is actually "**I WANT** but **I CAN'T**"! You feel that you are **DEPRIVING** yourself of something **DESIRABLE** and that battle is **EXHAUSTING**. And let's face it, the **CRAVINGS** usually **WIN**, don't they? That's **ADDICTION**!

I'll delve into this more later, but for now let's take a quick look at how **ADDICTION** works.

First, let's introduce the **LITTLE MONSTER**.

He was created by your first taste of **BAD SUGAR**, because that taste, and all the tastes of **BAD SUGAR** that you have had since, causes you to **MISS** it when you haven't had it for a while.

This is when the **LITTLE MONSTER** comes into play. He gives you that very subtle, empty, slightly insecure feeling that has you reaching for **MORE** – because the only way to keep him quiet is to keep up your dose of **BAD SUGAR**.

YUM! GIMME MORE.

SUGAR

12

Next, we have the **BIG MONSTER**.

He got to be **BIG** because he was fed by a number of sources – **ADVERTISING**, our **SOCIAL CIRCLES**, and even our **FAMILIES**.

This is the **BRAINWASHING** that tells you that having **BAD SUGAR** is some sort of **TREAT**, **A BOOST**, or **REWARD** that you can't live without.

And, indeed, you do get a temporary **BOOST**. The **LITTLE MONSTER** shuts up and you feel **BETTER** when he does – just like a junkie does when he has had his **FIX**. Until the **NEXT TIME** he needs a **FIX** and the **NEXT**…

In fact, you have been **ADDICTED** to sugar from the **VERY START**!

But you've got to ENJOY your food!

I **AGREE**! So, what if you could:

- **EAT AS MUCH OF YOUR FAVOURITE FOODS AS YOU WANT, WHEN YOU WANT?**

- **BE AT YOUR IDEAL WEIGHT WITHOUT DIETING OR SPECIAL EXERCISE?**

- **DO THIS WITHOUT USING WILLPOWER OR FEELING DEPRIVED?**

With the **PROVEN** and **EFFECTIVE** *Easyway* method, you **CAN**!

So do you want to **FREE** yourself from the slavery of **SUGAR ADDICTION** the **EASY WAY**?

Sure – but a LITTLE sugar now and then isn't a BIG DEAL, is it?

OK, here is the **BOTTOM LINE**...

THERE IS NO "HEALTHY" AMOUNT OF BAD SUGAR!

Now you're SCARING me!

14

Do you think that a **JUNKIE** having a **BIT** of heroin, an **ALCOHOLIC** having a drink **NOW** and **THEN**, or a **SMOKER** having an **OCCASIONAL** cigarette, is going to stop them from wanting and having **MORE**?

'A little bit' leads to **MORE** and **MORE** and the **ADDICTION** stays in place.

But people STRUGGLE to give up ADDICTIONS! It's HARD!

It's only **HARD** if you think **GOING WITHOUT** is hard!

But when you realize that sugar is doing **NOTHING** for you, it becomes **EASY**!

The only "sacrifices" are the ones you're already making now – through **POOR HEALTH, DISEASE,** and **SHORTENED LIFE EXPECTANCY**!

You'll get rid of all that when you're **FREE**!

Don't **WORRY** – I'll walk you through it and by the end of this book, you'll be **HAPPILY SUGAR-FREE!**

For now, you need do **NOTHING** except:

1. KEEP AN OPEN MIND
(*Easyway* has proven effective for all types of addictions since 1983. It is **TRIED** and **TESTED**, i.e. it **WORKS!**)

2. READ THE ENTIRE BOOK

3. DON'T SKIP ANY STEPS

And remember – it's **EASY!** That's why I called it *Easyway*!

IT'S ONLY NATURAL

OVEREATING is a significant problem in society. We can learn a lot from the **ANIMAL KINGDOM**.

ANIMALS DO NOT OVEREAT!

This is for NOW...   ...and this is for WINTER

There are only **THREE EXCEPTIONS:**

ANIMALS FED BY HUMANS

HUMANS

WILD ANIMALS THAT EAT HUMAN WASTE

17

So every other species eats only as much of their favourite foods as they want, as often as they want, without being **OVERWEIGHT**.

Do they use **WILLPOWER**? Do they go on **DIETS**?

OK, don't come BEGGING when you've RUN OUT!

Of course not! They **INSTINCTIVELY** know what they should and shouldn't eat.

And so did **WE**, once upon a time!

We still have this gift of **INSTINCT**, but refined sugar mimics the healthy sugar found in our **NATURAL FOODS**, i.e. fruit. This has caused **ADDICTION** to **BAD SUGAR**.

Hmmm...

But sugar comes from a PLANT! Isn't that NATURAL?

It may **START OFF** as **NATURAL**, but the **REFINING PROCESS** strips away **FIBRE**, **VITAMINS**, and **MINERALS**, leaving only 'EMPTY CARBS', with little nutritional value but more carbs than we can **BURN**. This excess is turned into **FAT**.

THE ILLUSTRATED <u>Easyway</u>

REFINED SUGAR causes chaotic false **HIGHS**, then **CRASHES** in **BLOOD SUGAR**.

We seek the **HIGHS**, but need **MORE** and **MORE** fixes of sugar to satisfy us when in reality, because sugar contains no nutritional value, it does not satisfy **HUNGER**. This creates a need to eat **MORE BAD SUGAR**.

Send a bowl of **FRUIT** around an office and you'll see people eat **ONE** piece, enjoy it, and leave the rest. Send around **SWEETS** and they will **GORGE** themselves!

BAD SUGAR has managed to slip under the radar in terms of **HEALTH EDUCATION** until now, mainly because almost everyone **CONSUMES** it.

However, the huge rise in **PREVENTABLE** and **REVERSIBLE** diseases, such as **TYPE 2 DIABETES** with its direct links to our sugar consumption, is too great to ignore. **DIABETES** has reached **EPIDEMIC** proportions with over **400 million cases** worldwide, a figure that is expected to increase by **50%** in the next **ten years**.

OVEREATING is a result of **INCORRECT EATING.**

By eating the wrong **TYPES** of foods, we tend to feel **HUNGRIER.** This is because the wrong foods do not contain the necessary **NUTRIENTS** to satisfy hunger. As a result, we eat **MORE** and the fat piles on!

The main culprit, of course, is **BAD SUGAR.**

Well, I don't have a SWEET TOOTH!! I prefer SAVOURY SNACKS!

PIZZAS, CRISPS, SAUCES, and **PRE-PACKAGED** and **READY-MADE,** or **TAKE-AWAY** foods typically contain **SIGNIFICANT** amounts of **BAD SUGAR.**

But I cook from SCRATCH!

It's not just refined sugar, but **PROCESSED, STARCHY CARBS** such as **POTATO, PASTA, RICE,** and **BREAD,** that are classed as **BAD SUGAR!**

ALL WHEAT products are also highly **PROCESSED** and thus classed as **BAD SUGAR.**

TRY THIS EXERCISE THE NEXT TIME YOU GO SHOPPING:

Pick up a **BASKET**   **AND a TROLLEY**

Now check the labels of every food item for **BAD SUGAR** and put them in the **BASKET** and **NON-BAD SUGAR ITEMS** in the **TROLLEY**.

NOW see if you're **ADDICTED** to **SUGAR**!

If the items containing **BAD SUGAR** were removed from supermarkets, only **20%** of stock would remain!

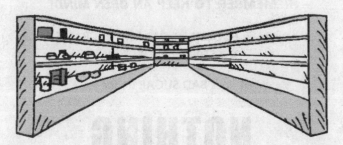

21

Adding **SUGAR** feeds the **ADDICTION**; poor-quality, tasteless items with no nutritional value are packaged as "food". Because **THEY'RE ADDICTIVE**, we **CRAVE** more, which means we **BUY** more. It's a **MARKETING STRATEGY** to keep us **HOOKED**. Nothing more!

IT'S TIME TO SEE BAD SUGAR FOR WHAT IT IS –

A BLAND SUBSTANCE THAT THROWS YOUR METABOLISM INTO CHAOS AND DOES NOTHING FOR YOU WHATSOEVER

AND

REMEMBER TO KEEP *AN OPEN MIND*!

You have been **FOOLED** into thinking that **BAD SUGAR** does something for you.

What does **BAD SUGAR** do for you?

NOTHING

THE
BRAINWASHING

Wow, she sure LOVES her FOOD!

Does she **REALLY**? Does **EATING** really make her **HAPPY**?

People who are **OVERWEIGHT** know full well that they **OVEREAT** and, at best, usually feel **GUILTY** and, at worst, can fall into **SELF-LOATHING**.

The person who eats the whole packet of **BISCUITS**…   …or others' **LEFTOVER BIRTHDAY CAKE**…

I'll just have ONE…

…or the secret stash of **CHOCOLATES**…

 …ends up feeling **BLOATED, DISSATISFIED,** and **ASHAMED**, which is far removed from **ENJOYING**, let alone **LOVING**, the **EXPERIENCE**!

Overeaters shouldn't feel **ASHAMED**. They were simply **BRAINWASHED** into being addicted! It's a myth that **WHAT** we eat and the **WAY** we eat it is a **FREE CHOICE**. Our food was **CONTROLLED** for us even before birth.

I've read all the food labels and now it seems as though my only option is STARVATION!

Though cutting out processed foods that contain **BAD SUGAR, ADDITIVES,** or **STARCHY CARBS** may seem daunting at first, when you begin to explore **HEALTHY OPTIONS**, you'll see how much **VARIETY** there is, probably more than in your current diet!

A FEW THINGS TO GET STRAIGHT:

YOU INSTINCTIVELY KNOW IF WHAT YOU ARE EATING IS NOT GOOD FOR YOU.

AND UNTIL NOW YOU HAVEN'T KNOWN HOW TO CHANGE.

BUT YOU ARE NOW WILLING TO TAKE CHARGE, ALTER YOUR MINDSET, AND CHANGE A SITUATION YOU'RE UNHAPPY WITH.

AND YOU GET IT THAT NO ONE IS IMPOSING THIS UPON YOU – THAT YOU ARE THE ONE IN CONTROL OF THIS CHOICE?

Yes!

AND THAT YOU ARE MAKING THIS CHOICE FOR AN EXCELLENT REASON?

Yes – to ENJOY life more!

So why would there be any **DOOM** and **GLOOM** about setting yourself **FREE** from the very thing that's **RUINING** your enjoyment of life?

Here's the GOOD NEWS!

Using the *Easyway* method:

You will not be on a DIET
You can eat foods you LIKE
You don't need to do an EXERCISE PROGRAMME
You don't need WILLPOWER

For the first time in your life, you will be genuinely CHOOSING your own food...

AND LOVING

WHAT YOU EAT!

I'm not CONVINCED about your idea of "favourite" food!

If you follow the method, you'll find that your "**FAVOURITES**" will have changed.

You're not being **CONNED**; you're being **ENLIGHTENED**.

You've **OPENED** your **MIND** and now see the **TRUTH**: that you didn't actually **LOVE** the food you ate!

OK – have your current **FAVOURITES** made you **HAPPY**?

I'm not sure about THAT!

Um, no, not REALLY...

What kind of favourite food makes you **UNHAPPY**?

Well the unhappiness comes later – AFTER the WONDERFUL TASTE is finished!

It's actually the way you **THINK** about how something tastes that determines how it tastes to you!

Huh?

The first taste of many **DELICACIES** is usually quite **REVOLTING**!

But because we think of them as **DESIRABLE LUXURIES**, we become convinced that the taste is **PLEASURABLE**.

Did you know that the cake you're eating is made with processed **CHICKEN FEATHERS** and **RAT'S MILK**?
(It's a way the big bakeries get cheaper ingredients.)

PTOO!

Of course **I'M KIDDING**, but did you notice how the **IDEA** of what you were eating affected the way it **TASTED** to you?

Don't worry – you're not going to have to picture **RAT'S MILK** for the rest of your life to deter you from eating cake!

But the way you **THINK** something tastes is influenced by **BRAINWASHING** and that **BRAINWASHING** began the day you were born.

It was **DECIDED** for you when you would:

BE WEANED   **EAT SOLIDS**   **EAT CERTAIN FOODS**

IN FACT, YOU WERE ABSORBING BAD SUGAR BEFORE YOU WERE EVEN BORN

31

Even as an **ADULT**, your **CHOICES** have been **LIMITED** by: **EARLY CONDITIONING, SOCIAL EXPECTATIONS,** and **ADVERTISING**.

Your eating habits are a result of **EXTERNAL INFLUENCES** and **LIMITED CHOICE**.

The result is a culture of **BRAINWASHING** that has become so established, we don't even see it unless it is pointed out to us.

Aren't the SNACKS between meals the PROBLEM? Isn't that just IMPULSE buying and therefore FREE WILL?

Absolutely **NOT**. Your desire for these has been triggered by **BRAINWASHING**.

Advertisers have only **ONE GOAL** – to get you to buy **MORE**, so they set out to convince you that the **SUGARY** treat is an "**ENERGY**" bar, or that it adds to your **KUDOS**, or makes you **SEXIER**!

BRAINWASHING also controls **HOW MUCH** we eat.

33

And as for **CHRISTMAS**....

...it's an **ORGY** of **EATING!** You'd think feeling so **SICK** and **BLOATED** from such excessive gorging would be enough to make you want never to repeat it and yet we're back for more **YEAR** after **YEAR**.

And for some people, it's like Christmas **ALL YEAR!**

So why do we OVEREAT?

Obviously the **DISCOMFORT** of overeating is not enough to put us off. Something else is at play and that is **ADDICTION**.

With other addictions, you have to **WORK** at becoming addicted. Most smokers don't like the **TASTE** of **CIGARETTES** at first. They have to **PUSH THROUGH** the unpleasantness until they develop a "**TASTE**" for it.

But what they actually do is **LOSE** their sense of taste. The need for the substance over-rides all of their natural **WARNING INSTINCTS**.

> But wait a minute! How does that apply to **FOOD**? I've always **ENJOYED** eating!

Unlike other **ADDICTIVE SUBSTANCES**, we become hooked on sugar before we even **REALIZE** we're eating it, so there's no period of **ADJUSTMENT**.

By the time we reach **ADULTHOOD**, we're **CONVINCED** that **BAD SUGAR** products give us some **BENEFIT**, even though all we feel is **BLOATED**, yet still **UNSATISFIED**, so we eat **MORE**.

This is **FALSE HUNGER**.

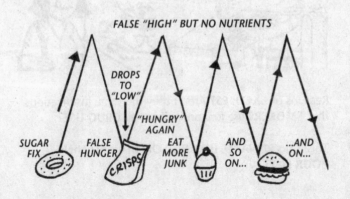

FALSE "HIGH" BUT NO NUTRIENTS

SUGAR FIX

FALSE HUNGER

DROPS TO "LOW"

"HUNGRY" AGAIN

EAT MORE JUNK

AND SO ON...

...AND ON...

Let's explore the **REASONS** people give for eating **JUNK**:

Reasons or **EXCUSES**? Aren't they the same justifications that **SMOKERS** use for continuing their **DRUG USE**?

This is no **REWARD**. It's a **POISON** that's **RUINING YOUR LIFE!**

But **SOMETHING WONDERFUL** is about to happen!

You are going to be **FREE** of:

FEELING BLOATED INDIGESTION HEARTBURN
GUILT SHAME FEELING LIKE A FAILURE BINGEING
CRAVINGS COUNTING CALORIES DIETING

And you will **GAIN**:

PERMANENT WEIGHT LOSS HEALTH
VITALITY ENERGY ENJOYMENT OF FOOD
CONTROL OF YOUR LIFE SELF-CONFIDENCE!

I'm sorry but I find this TOO GOOD to be TRUE! I've tried every DIET there is and ended up having my HOPES CRUSHED!

The *Easyway* method does **NOT** ask you to **DIET**!

DIETING involves **WILLPOWER** and trying to exercise that for a **LIFETIME** just sets you up to **FAIL**.

You are not being asked to **SACRIFICE, RESTRICT,** or **DEPRIVE** yourself as **DIETING** demands. Any failures you have experienced in the past are not because of a weakness in **YOU** but in the **METHOD** you used.

Easyway is **TRIED** and **TESTED** – it **WORKS** and it's **EASY!**

All you need do is **FOLLOW THE INSTRUCTIONS** and here's the next one:

**START OFF WITH A FEELING OF
EXCITEMENT AND ELATION**

You're about to be set
FREE!

THE TRAP

I've tried to just go COLD TURKEY on BAD SUGAR and even though it had left my SYSTEM, I still CRAVED it! Why?

ADDICTION isn't just **PHYSICAL!** In fact, **99%** of **ADDICTION** lies in the **MENTAL** realm!

The problem lies in the **BELIEF** that you derive some **PLEASURE** or **CRUTCH** from eating **BAD SUGAR** foods.

As long as you believe that, you'll feel **DEPRIVED** without them.

Remember the
LITTLE MONSTER?
He belongs to the
PHYSICAL REALM.

You can **KILL** him off by
going **COLD TURKEY**, simply
by cutting off his supply of
the addictive substance.

But **HE'S** not the
REAL problem!

> NO MORE for you!

> FEED ME!

> NO, NO NO!
> You NEED this!

Unless you also
kill off the **BIG
MONSTER** in
the **MENTAL
REALM** – the
BRAINWASHING
that tells you
this thing **DOES
SOMETHING** for
you – you'll never
be **FREE** of the
DESIRE for it.

Here's how it **WORKS:**

YOU HAVE A CRAVING FOR CAKE AND SEEK ONE OUT.

ONCE YOU HAVE IT, YOU FEEL IMMEDIATE RELIEF.

BUT IT HASN'T EVEN REACHED YOUR STOMACH YET!

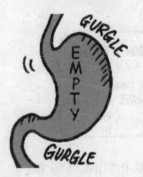

The **RELIEF** you feel has **NOTHING** to do with the **FOOD** and satisfying the Little Monster's **HUNGER PANGS.**

It has **EVERYTHING** to do with feeding the Big Monster's **ADDICTION.**

Let's take a look at how we become **TRAPPED** in **ADDICTION** by observing nature's pitcher plant and how it traps a fly.

Having been **EXPOSED** to **SUGAR**, you become **SEDUCED** by it.

You **IGNORE** all your **NATURAL INSTINCTS** for the "**PLEASURE**" of this "**REWARD**".

Until you find yourself completely **TRAPPED!**

The difference between **SUGAR ADDICTION** and other addictions is that we find ourselves in the pitcher plant, not through our own **CHOICE** but through the choices that have been made **FOR** us.

Before **BIRTH**, we started consuming **BAD SUGAR**:

- The **BAD SUGAR** our mothers ate while **PREGNANT**

- Followed by **BABY FOOD** containing bad sugar

- **BABY FOOD** containing highly processed **FRUIT SUGAR**

- **STARCHY CARBS** such as **CEREALS, POTATO, RICE, PASTA**, and **BREAD**

- "**TREATS**" such as **CAKES, BISCUITS, CONFECTIONS, CRISPS**, and **SUGARY DRINKS**

43

So by the time we reach **ADULTHOOD**, we have

ALREADY LOST CONTROL!

The **HUMAN BODY** is an **INCREDIBLE, SELF-CORRECTING, ADAPTABLE** machine.

When you consume something that is **BAD** for you, it **GETS RID** of it for you!

But if you **KEEP** feeding yourself **POISON**, your body builds up a **TOLERANCE** to the poison, so it takes **MORE** to have an effect.

So if you want to get the same "**HIGH**" from the substance, you need **MORE** and **MORE** and **MORE**.

At the same time, the **DOUBLE LOW** of physical and mental **CRAVING** in between highs becomes more **EXTREME**.

Therefore the **ILLUSION** of "pleasure" when you relieve the cravings makes the drug (food) seem more **PRECIOUS**, and so it goes **ON** and **ON** and **ON** until **NO AMOUNT** of **FOOD** will **EVER** be **ENOUGH** to satisfy you and you stray further and further from the "**HIGH**" you seek.

If I just have a LITTLE BIT MORE, I'll get some RELIEF!

While the **BIG MONSTER** tells you that **BAD SUGAR** makes you **HAPPY**, in your heart of hearts you know that this just isn't **TRUE**, don't you?

Yes, that's why I'm reading this BOOK!

The only way to reverse the inevitable decline is to

COMPLETELY STOP EATING BAD SUGAR.

Sorry, but I find that idea a bit DAUNTING!

So the question is
"What's STOPPING you?"

Ummm...

46

What's **STOPPING YOU** is that you still believe in a few stubborn **MYTHS**. So let's **EXPLODE** them now.

MYTH 1: SUGAR, STARCHY CARBS, or PROCESSED CARBS constitute a **CRUTCH** or give you **PLEASURE**.

When you come to think about it:

It's not the **FOOD** that brings you **PLEASURE**!

It's the **COMPANY**, the **SETTING**, and the **OCCASION**.

Take away the **FOOD** and you can still **ENJOY** the occasion.

Take away the **COMPANY** and it won't hold much **MEANING**.

MYTH 2: ESCAPE WILL BE HARD

You're not **AVOIDING FOOD**, you're changing to **REAL FOOD** and when you open your eyes and look around, you'll find there's so much of that to **ENJOY**, instead of **HEAVY, FATTY, STODGY, CHEMICAL-FILLED JUNK!** And enjoy it you will, because you'll be feeling so much **BETTER** about yourself and life.

The food you eat reflects how you **FEEL:**

LIFELESS AND HEAVY   **VITAL AND FRESH**

Now let's get out the **BIG GUNS!**

It's time to tackle the **BIG MONSTER!**

STARVING THE BIG MONSTER

You've already come a long way to understanding how to **FREE** yourself from the trap of **SUGAR ADDICTION**, but there is no "**MAGIC SECRET**" you are yet to discover.

You just need to really "**GET IT**" that you're not **SACRIFICING** anything and that is what this **ENTIRE BOOK** is about.

So let's look closely at the **BIG MONSTER** which is the **BRAINWASHING** that convinces you that you simply can't live without **BAD SUGAR!**

Most people who are trying to free themselves from **ADDICTIONS** and those who try **DIETING** tend to **FAIL** because they are using **WILLPOWER** alone, without removing the **DESIRE** for that which they **CRAVE**. They find themselves in a **TUG-OF-WAR**.

> I KNOW it's KILLING me, but I'd BE MISERABLE without it!

> I KNOW how BAD it is for me, but it's like those warnings on cigarette packs which don't work. I'm wondering how I could cope without it.

> SMOKE KILLS

This **TUG-OF-WAR** seems to be between the knowledge that the substance is **BAD** for you and the belief that life would be **TERRIBLE** without it.

However, once you realize that the only battle is between your choosing between a **FACT** – that bad sugar does **NOTHING** for you – and a **MYTH** – that it provides some **PLEASURE** or **CRUTCH** – you're **HOME FREE!**

This is **CRUCIAL** – if you keep thinking that **BAD SUGAR** is **DESIRABLE**, you'll feel **DEPRIVATION** when you stop and you'll be back to using **WILLPOWER** to fight that **DESIRE**, which will keep you at **RISK** of falling back into the **TRAP** when the **WILLPOWER** slips.

Consider this: **YOU ARE IN EXACTLY THE SAME TRAP AS ANY ADDICT!** Look at the "logic" that keeps these people **TRAPPED** – aren't you using the same **ARGUMENTS** about **SUGAR?**

ARE they **REALLY** so **TASTY**? Let's see…

Would you **ENJOY** eating a bowl of **PLAIN RICE** or… **PASTA** on its **OWN**?

Would you eat a **TABLESPOON** of **SUGAR** on its **OWN**?

52

How much **BREAD** could you eat without adding a **SPREAD**?

How about **PASTRY** on its **OWN**?

If these foods are so GREAT, why can't they simply be enjoyed on their own?

We always need to add REAL FOOD to make them appealing and NUTRITIOUS.

FRUIT and **VEGETABLES*** can be eaten **RAW** and are perfectly **SATISFYING** without needing **EXTRAS** to make them **PALATABLE**.

But ready-made food is just more CONVENIENT!

You may as well be eating the **PACKAGING** if you're seeking **NUTRIENTS!**

The manufacturers are not interested in your **CONVENIENCE!** They just want you to eat more of their **SUGARY PRODUCTS!**

Ask yourself: am I really **HUNGRY** or am I just being tempted by the **DISPLAY?**

* The **EXCEPTION** to this is **POTATOES** – which are really **STARCHY** which is why they are to be excluded. You would never eat a **RAW** potato by choice, would you?

There's nothing "natural" about eating a potato.

BAD SUGAR SNACKS	FRUIT
Bad for you	**Good for you**
Little nutritional value	**Highly nutritious**
Cost more overall	**Inexpensive**
Packaging creates waste	**Biodegradable**
Don't satisfy hunger – you	**"packaging"**
always want more and overeat	**Satisfies hunger**

Are you starting to see the **TRUTH** about **BAD SUGAR**?
Or do you still have some beliefs that are keeping you
ATTACHED to it?

Let's explode some more **MYTHS**!

TRUST YOUR GUT FEELINGS

Have you ever come across a more **INCREDIBLE MACHINE** than the **HUMAN BODY**?

No matter what you **THROW** at it, it constantly seeks to:

- **REBALANCE**
- **ADJUST**
- **COMPENSATE**
- **ADAPT**
- and **SURVIVE**.

And like another **AMAZING MACHINE** – the **CAR** – it will run **PERFECTLY** well for many years, given **CARE, MAINTENANCE,** and the **RIGHT FUEL**.

There are quite a few things that a **HUMAN BEING** and a **CAR** share and the most important, when it comes to **HEALTH** and **WELLBEING**, are **WARNING SIGNS**. Ignore these and you're in **TROUBLE!**

But unlike the car, our **INCREDIBLE MACHINE** has a **UNIQUE**, **ADDITIONAL**, and, sadly, often **NEGLECTED** gift:

And it is the **IGNORING** of this that is the **FLAW** in our otherwise **PERFECT** machine.

If only we had listened more to "**NATURE'S GUIDE**", there would be no need for this book! "**NATURE'S GUIDE**" will reliably tell you what you should or shouldn't **EAT**.

Ever known a child to **THROW UP** after gorging on sugary food at a birthday party?

For some reason, we shrug this off as **NORMAL**!

It's **NOT**! A child's body is not yet **CONDITIONED** to sugar. It knows it's a **POISON** and purges it to **PROTECT** her.

Nor is **INDIGESTION** normal if we are eating a **HEALTHY DIET**. There is no reason for your body to react negatively to something that is **GOOD** for it.

Indigestion is telling you that something you ate is **BAD FOR YOU**.

What do we do?

We **IGNORE** these signals and carry on or we **MASK** the discomfort or pain with **DRUGS** – many of which actually contain **BAD SUGAR**! – without removing the very thing that gave us the problem in the **FIRST PLACE!**

Like **BAD SUGAR**, drugs are **POISONS** in controlled doses.

And, like sugar, your body **ADAPTS** to the poison, and so you need **MORE** to get the same effect.

Meanwhile, the **REAL PROBLEM** remains unresolved, while your body has an **EXTRA BURDEN** of alien chemicals to deal with!

If the **OIL LIGHT** on your car was **FLASHING**, would you expect it to **FIX** itself if you put tape over the light so it didn't **DISTURB** you?

Taking **MEDICINES** to alleviate **SYMPTOMS** without addressing the **CAUSE** is doing the same thing!

NATURE'S GUIDE is your inbuilt **WARNING SYSTEM**. If you **IGNORE** it without fixing the **REAL PROBLEM**, your **BODY**, like your **CAR**, will eventually **BREAK DOWN**.

Wow! I'm starting to see what you mean! It's a real MESS. But how did we lose sight of NATURE'S GUIDE to such a degree?

As we have seen, we become **CONVINCED** that we **NEED** to eat bad sugar and "empty carbs" through…

INVOLUNTARY EARLY EXPOSURE TO SUGAR

CONDITIONING

ADVERTISING

and MISINFORMATION.

But look at the **ANIMAL KINGDOM**. Big strong animals don't need to consume milk, or dairy products, beyond infancy to **STAY HEALTHY**. It's the same for fast, strong, muscular animals – feed them **PASTA** and they'd soon look bloated, overweight, and unfit. And they'd also start feeling **SICK**!

Hmm - this will fill a gap!!

Being the **INCREDIBLE** (and obedient) **MACHINE** that it is, your **BODY** does its best to **ADAPT** to these unnatural foods.

It can only work with what it's **GIVEN**, so it does the **BEST** it can with less than **IDEAL** sustenance.

And, historically, in times of **SCARCITY**, the body had to adapt and could be **TRICKED** into developing a taste for food that was **AVAILABLE** until better food was restored, but it was **NEVER** meant for us to **KEEP** eating the sub-standard food!

But in modern times, we can lengthen the life of food. **HOW**? Through **PRESERVATIVES**.

PRESERVING food through **COOKING, SALTING, FREEZING, PICKLING, BOTTLING, CANNING**, and **REFINING** may protect food from the **BACTERIA** that cause it to **DECAY**, but it also strips it of **KEY NUTRIENTS**. And **BAD SUGAR** is often used as a **PRESERVATIVE**!

Your body acquires a **TASTE** for these foods by two means:

1. OVER-RIDING THE WARNING SIGNALS

2. SUPPRESSING THE NATURAL REFLEX TO REJECT THESE FOODS.

By **FORCING** the body to adapt, you destroy its ability to **PROTECT** you and you lose sight of your inner guidance – your **INSTINCT**.

HOWEVER, just as you were **CONDITIONED** to crave second-rate foods, your **INSTINCT** is still well and truly **INTACT** and can be **RESTORED**.

Once you break free of the **BRAINWASHING**, you'll rediscover the **PLEASURE** of eating healthy, life-giving food.

It's time to **FREE YOURSELF FROM THE SLAVERY OF EATING HABITS THAT ARE KILLING YOU! TRUST YOUR INSTINCT!**

WHAT GOES IN MUST COME OUT

Many people use **THESE** as the gauge of their **FITNESS**...

...when really it's **THIS**...

and **THESE**!

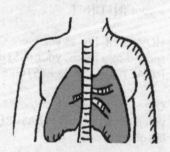

If you **LIKE** what you **SEE** in the mirror and can carry out your daily activities without gasping for **BREATH**, why does it **MATTER** what the **SCALES** say?

If you're **HAPPY**, you're **HAPPY**... and **THAT'S** the **IDEAL**!

SO HERE'S YOUR NEXT INSTRUCTION:

DISREGARD ANY PRECONCEIVED WEIGHT TARGET!

Besides, our goal is to get you off **BAD SUGAR** – the rest: good health, the ideal weight for you, and fitness – are just **BONUSES**.

So WHY do we PUT ON weight?

You gain weight when your **INPUT** of energy (food) is greater than your **OUTPUT** of energy (activity).

While these conditions may have some **INFLUENCE** over your ability to lose weight, the formula remains the same:

IF YOU PUT IN MORE THAN YOU PASS OUT, YOU'LL GAIN WEIGHT!

Do you still eat **JUNK**? And do you think you can eat **MORE BECAUSE** you exercise?

Exercise is **GREAT** – keep doing it – but the same rule applies: if you put in more than you eliminate, you'll **GAIN WEIGHT**!

It's not even a case of calories. It's what kind of calories you're putting in that matters.

IT'S IMPOSSIBLE TO OVEREAT NATURAL FOOD

OK then – what if I reduce my INTAKE OF ALL FOOD?

If you do so, you'll be **DIETING** and **DIETING** requires **WILLPOWER**.

How **LONG** can you keep up that **STRUGGLE** with yourself? Eventually, you'll revert to form.

Well, if EXERCISE and DIETING don't work, what DOES?

Eating the **RIGHT FOODS**! When you do, you can eat **AS MUCH** of your **FAVOURITES** as you like and both gain and maintain an ideal weight!

Here's a question: **WHY DO WE EAT?**

To avoid **DYING**, of course!

Is that the **REAL** reason?

You might often say:

I'm STARVING!

GURGLE

but the reality is that you're nowhere **NEAR** it, so eating for **SURVIVAL** isn't the issue.

We often eat for the following reasons:

ROUTINE **REWARD OR COMFORT** **BOREDOM**

LUNCHTIME!

But none of these is the **REAL REASON** we eat! The **REAL** reason we eat (if we're running on **INSTINCT**) is because we feel **HUNGRY**!

TRUE HUNGER is triggered by a fall in **NUTRIENTS**, which signals that it's time to eat. It is **NATURAL** and not **UNPLEASANT** unless...

YOU HAVE NO ACCESS TO FOOD, OR...

...YOU ARE STARVING YOURSELF ON A DIET!

HUNGER is related to **TASTE** – the hungrier you feel, the better food tastes to you. That's how our bodies can **ACCEPT** and even **CRAVE** second- or third-rate foods in times of **SCARCITY**.

You'll eat **ANYTHING** if you're **HUNGRY** enough!

SURVIVAL SHOW

If you eat when you're not **GENUINELY** hungry, food will be less **SATISFYING**, so you will eat **MORE**.

As we have seen, hunger signals a drop in **NUTRIENTS**.

If the food you eat doesn't contain those nutrients, your hunger won't be **SATISFIED** and therefore won't be **SWITCHED OFF**.

HUNGRY, HUNGRY, HUNGRY!

This is **FALSE HUNGER**, which **BAD SUGAR** cannot satisfy! The **BALANCE** between **INTAKE** and **OUTPUT** will always be out when you try to gain nutrients from **BAD SUGAR** products.

The more **EFFICIENTLY** you satisfy hunger (with **NUTRIENT-RICH, NATURAL, UNPROCESSED FOODS**), the less you will need to **CONSUME**!

So here is your next instruction:

AVOID EATING WHEN YOU'RE NOT HUNGRY

FACING DOWN YOUR FEARS

You'd think that giving up a **POISON** that really has no **TASTE** and gives you few **NUTRIENTS** and leads to so much **ILLNESS** would be **EASY**, wouldn't you?

Well guess **WHAT**? It **IS** – once you get the **FEARS** out of the way!

So YOU say!

So you **STILL** think that bad sugar gives you some sort of **PLEASURE** or **CRUTCH**?

Well, it MUST do if I WANT it so MUCH!

Well, it's now time to address the thing that most keeps you in the **PRISON** of **ADDICTION:**

72

FEAR is a response that comes from **INTELLECT** and **INSTINCT**. It is **HELPFUL** in ensuring our **SURVIVAL** in the face of **DANGER**.

YIKES!

And it can urge you to **HOLD ON TO** the things that ensure your **SURVIVAL**.

I need this JOB!

But what if the things you **FEAR** are based on **ILLUSIONS**? It's time to **SHATTER** those!

So what do you **FEAR** about giving up **BAD SUGAR**?

So you won't eat that **CAKE** in the fridge if there's no one to **SHARE** it with?

The only benefit about **SHARING** an addiction is that it makes you feel better about **YOURS**...

...that is, for a **WHILE** – until **GUILT** and **REMORSE** set in.

You'll never be **RELAXED** as long as **THE LITTLE MONSTER** keeps pestering you to **FEED** him his **SUGAR HIT!**

Well, OK then – it gives me ENERGY!

Using **SUGAR** for an energy hit is like getting a **PAYDAY LOAN**: it's a small, **TEMPORARY** high that keeps you in **DEBT** because you soon feel **EXHAUSTED** and need **MORE** and **MORE** to keep going.

I love the SMELL!

Does that mean you're compelled to drink **PERFUME**? Or to eat **FLOWERS**?

Look at **WILD ANIMALS**.

Untameable, muscular, perfectly equipped to run, fight, and react to anything life throws at them.

They don't need **BAD SUGAR FIXES** to keep them that way. **BAD SUGAR** doesn't help us either.

It's my COMFORT FOOD and a TREAT or REWARD!

If you're **UPSET**, you eat **SUGARY FOOD**.

But when you're **HAPPY**, you eat **SUGARY FOOD** too! When you **FAIL**, when you **WIN**; when you're **BUSY**, **IDLE**, **UP**, **DOWN**, **IN** or **OUT**, you eat **SUGARY FOOD**!

The **TRUTH** is you're **ADDICTED** to **SUGARY FOOD**!

And there's **ABSOLUTELY** no "reward" or "comfort" in feeling **ASHAMED** and **DISGUSTED** with yourself!

Do you even **TASTE** the food you're **WOLFING** down?

But I LOVE the TASTE!

If it was really all about **TASTE**, why don't manufacturers make **SUGAR-FREE** chocolates?

Because then you wouldn't want **MORE**, would you?

And – if you actually think that **CHOCOLATE** is better than **SEX**, you need **HELP**!

ADDICTS love being with other **ADDICTS** because it offsets their own feelings of **GUILT** and **INADEQUACY**.

> Life without sugar would be BORING! I'm more FUN and LOVEABLE when I'm "SUGARED UP"!

They'll even **URGE** you to have **MORE**, because it makes them feel better about **THEMSELVES** – not because it makes you more **ENDEARING** to them!

Trying and failing using other methods can be **CRUSHING**, but do you think that if you never again **TRY** to **ESCAPE**, you're more likely to make escape **POSSIBLE**?

> But what if I TRY and FAIL?

Could you be any **MORE MISERABLE** than you are now – trapped in **SLAVERY**?

Well, I'm afraid that if I SUCCEED, I'll never stop being MISERABLE!

Remember the **TUG-OF-WAR** between **HATING** the fact that you're addicted to **BAD SUGAR** but **FEARING** life without it?

Well, the only thing pulling you from **BOTH SIDES** is **BAD SUGAR**!

Eliminate **BAD SUGAR** and the **TUG-OF-WAR** is **OVER**!

TAKE AWAY THE BAD SUGAR AND THE FEAR GOES TOO!

Let's get this **STRAIGHT**!:

You're in the **PRISON** of **ADDICTION** because **BAD SUGAR PUT YOU THERE!**

Your only **JAILER** is **BAD SUGAR**!

Get rid of **BAD SUGAR** and you're **FREE**!

You are not about to **SACRIFICE** anything!

All you are doing is **REMOVING** the very thing that is making you **MISERABLE** and replacing it with **GENUINE WELLBEING**, instead of the illusory "boosts" that you think sugar gives!

Those "boosts" are just the result of **TOPPING UP THE DRUG**, nothing more!

How long do those "boosts" last anyway, before you start to feel **BLOATED**, **GUILTY**, and **OUT OF SHAPE**?

The only way to stop those terrible **CRAVINGS** is to **REMOVE** the **DRUG** that's causing them – a drug that's doing **NOTHING** for you in the first place!

Imagine finishing every meal feeling **FABULOUS** instead!

Isn't that something to **LOOK FORWARD TO**?

You're not **LOSING** a **FRIEND**, you're **ELIMINATING** an **ENEMY**!

DITCH THOSE DOUBTS!

DIVE INTO YOUR NEW LIFE!

DITCH THE SUGAR NOW!

THERE IS NOTHING TO FEAR!

NO WILLPOWER NEEDED

ADDICTION makes us feel **HELPLESS**. We assume that our inability to quit is because of a **WEAKNESS** in our character, but more often than not the **OPPOSITE** is true.

> But I must be WEAK because I just can't RESIST! If only I had more WILLPOWER!

The assumption that quitting any addiction requires **WILLPOWER** is widespread.

> Just don't give in to TEMPTATION!

If **RESISTANCE** is so **EASY**, why do so many good, intelligent people stay **HOOKED**?

The only reason people fail is that they are using the **WRONG METHOD**. Any method involving **WILLPOWER** keeps you in **CONSTANT CONFLICT** with yourself.

EATING SUGAR IS HURTING ME...

BUT THE PAIN OF LIVING WITHOUT IT IS WORSE.

Say you were locked in a **PRISON CELL**. Someone tells you that, by **PUSHING** on a certain point, the heavy door will **OPEN**.

So you **PUSH** and **PUSH**... until your **STRENGTH** gives out and you're still **LOCKED IN**. You conclude that:

I'm not STRONG enough to break out of here!

I'll NEVER get out of here!

That's the **WILLPOWER METHOD**.

Now imagine that someone else comes along and points out that you've simply been pushing in the **WRONG SPOT**!

> See! It opens **EASILY** when you know **HOW**!

That's the Easyway method. Would you give it a **TRY** or stay **IMPRISONED**, believing there's no **EASY WAY** out?

With the **WILLPOWER METHOD**, you try to focus on all the reasons to stop and hope that you are **STRONG** enough to hold out until the temptation subsides. But how **LONG** will that take?

At the beginning, you may feel good about the **SACRIFICE** you are making, but soon that feeling turns to **RESENTMENT**, **IRRITABILITY**, and **MISERY**. You want to feel **BETTER** and there's one thing you think will help you do that...

YOU EAT BAD SUGAR!

Now you feel even **MORE** miserable!

> I'm so **WEAK**! And I don't feel **BETTER**, I feel **WORSE**! And now I'm **HOOKED** again!

BUT YOU ONLY NEED WILLPOWER IF YOU HAVE A CONFLICT OF WILL!

The *Easyway* method doesn't focus on the **NEGATIVES** of quitting, but helps you see all that you will **GAIN** when you do. As you come to see that there is nothing to **FEAR** and that you are losing **NOTHING**, you can start to **LOOK FORWARD** to a life without sugar, thus removing any **CONFLICT**.

I must be WEAK-WILLED to have become ADDICTED!

It takes a **STRONG WILL** to persist with something that goes against all your **NATURAL INSTINCTS!**

WEAK-WILLED? Let's see:

- **YOU'RE ORGANIZED** – you find strategies to avoid detection.

- **YOU'RE DEFIANT** – if people tell you to stop, don't you do the opposite?

- **YOU'RE PERSISTENT** – you've tried and failed to quit many times.

- **YOU'RE DETERMINED** – you put up with an enormous amount to stay addicted.

Think of all the **FAMOUS, SUCCESSFUL** people who are **OVERWEIGHT**, who **SMOKE**, take **DRUGS**, or **GAMBLE** – are they **ALL** weak-willed?

The **WILLPOWER METHOD** is like running a marathon in **TIGHT SHOES**.

You're determined to finish, but the further you go, the more painful it becomes and the **FEAR** of **FAILURE** becomes stronger.

The stronger your **WILL**, the longer you withstand the agony, but with **WILLPOWER** there **IS NO FINISH LINE**!

The _Easyway_ method is different because **YOU REMOVE THE TIGHT SHOES**! And there's no running: it's a walk in the park.

AAH, the RELIEF!

And beware the **BRAINWASHING** of those who try to quit using the **WILLPOWER METHOD** and reinforce the myth that quitting is **HARD** and involves **SACRIFICE**.

WHINERS   **BRAGGERS**

It's AWFUL!

It has been an HEROIC struggle to hang in there!

NEXT INSTRUCTION: IGNORE THE ADVICE OF ANYONE WHO CLAIMS TO HAVE QUIT BY WILLPOWER!

I'm starting to see what you mean and it makes SENSE, but I really don't think I'm WEAK-WILLED...

...so I think I must just have an ADDICTIVE PERSONALITY!

This is a convenient theory that keeps you in the **TRAP**. But it is just an **EXCUSE**. I'll now explain how it works.

ABOUT ADDICTION

Any character traits shared by addicts are not the **CAUSE** of their addiction; they are the **RESULT**.

Your struggles leave you feeling **CONFUSED** and **FOOLISH**.

I can't understand why I can't GIVE UP! I have control over so many other areas of my life!

Until you can clearly see how you've been **TRAPPED**, you'll think the problem lies with **YOU**. Then you start making **ILLOGICAL EXCUSES** to explain your **IRRATIONAL BEHAVIOUR**:

And if you concede you're **ADDICTED**, you always have this **FALLBACK**...

...which works on many levels to keep you **TRAPPED**.

The idea of having an "addictive personality" gives you an excuse to:

- **AVOID** the fear of **SUCCESS** by not even **TRYING**
- **CLOSE YOUR MIND** to the idea that it might be **EASY**
- Feel **SELF-PITY** instead of your usual **SELF-LOATHING**.

> But I've heard EXPERTS say that some people are GENETICALLY PREDISPOSED towards ADDICTION!

It's no more than a **THEORY**; largely based on the incidence of **MULTIPLE ADDICTIONS** in the same individuals or family groups, e.g. **SMOKING** coupled with **DRINKING**.

In fact, these addictions are **CAUSED** by something they have in common: **THE MISTAKEN BELIEF THAT THE ADDICTION GIVES YOU SOME KIND OF PLEASURE OR CRUTCH!**

> No, sorry, I've TRIED and FAILED too many times! It's beyond my POWER to do this! I must have some CHARACTER FLAW!

I can understand how you might come to that **CONCLUSION**, but **FEAR NOT**! You **CAN** do this! I'm going to **KEEP GOING** until you "**GET**" this and the **EVIL SPELL** is finally **BROKEN**!

It's time to revisit the **BIG MONSTER** and the **LITTLE MONSTER**.

The **LITTLE MONSTER** is the **ADDICTION** – that mild, slightly empty, restless feeling which comes when **BAD SUGAR** is withdrawing from your body.

SCRATCH SCRATCH

The **BIG MONSTER** is the **BELIEF** that bad sugar is **ENJOYABLE** and that the feeling you get when you relieve the **WITHDRAWAL** is genuine **PLEASURE** and the **LONGER** you wait for your **TOP-UP**, the **WORSE** the feeling of being **DEPRIVED** and the **GREATER** the feeling of **RELIEF**.

YOU NEED BAD SUGAR!

The **WILLPOWER METHOD** focuses only on killing the **LITTLE MONSTER**...

> If I stop feeding the **LITTLE MONSTER** long enough, the drug will be **GONE** and so will the **CRAVINGS!**

...and ignores the **BIG MONSTER**, which actually makes him **STRONGER** because you feel as though you are making a **SACRIFICE**.

As long as you allow the **BIG MONSTER** to live on in your head, you'll always feel **DEPRIVED** and will never be free of those **CRAVINGS**.

As long as the **BIG MONSTER** exists, he'll convince you that you **NEED** your **SUBSTANCE** to help you through life and especially its **EXTREMES**, **BIG EVENTS**, or **OCCASIONS** such as:

TRAUMA

BEREAVEMENT

SOCIAL MILESTONES

LIFE CHANGES

THE SO-CALLED ADDICTIVE PERSONALITY IS NOTHING MORE THAN THE BIG MONSTER AT LARGE

Killing the **BIG MONSTER** is easy if you give up the idea that you have an **ADDICTIVE PERSONALITY** or **GENE**. You just need to release yourself from the **BIG MONSTER'S BRAINWASHING**.

> Defeating TWO monsters sounds like a tremendous BATTLE to me!

It's not a matter of **BATTLING** these two, but instead clearly seeing how they have **MANIPULATED** you into believing that you can't do without **BAD SUGAR**.

Let's get a **CLEAR PICTURE** of how they work.

BOTH monsters were created by your first exposure to **BAD SUGAR** and work together, but in different ways.

FEED ME!

The **LITTLE MONSTER** works on a **PHYSICAL** level. He developed a **TASTE** for **BAD SUGAR** and now relies on you to keep up a **REGULAR** supply. If you don't give him what he wants, he lets you know by a small, subtle **ITCH** that tells you to **TOP UP** the dose.

The **BIG MONSTER** works on the **PSYCHOLOGICAL** level. He is the product of all the **CONDITIONING** you have been exposed to – even in your **EARLIEST** years – and this has you **CONVINCED** that **BAD SUGAR** gives you something.

THE LITTLE MONSTER GIVES YOU THE ITCH...

...BUT IT'S THE BIG MONSTER WHO CONVINCES YOU THAT YOU *MUST* SCRATCH IT.

But my gran only has ONE biscuit a day and my friend can eat a FEW chocolates and leave the rest!

That may be true, but you know that for you that's **IMPOSSIBLE.**

BUT ONCE YOU'RE FREE, YOU'LL WONDER HOW ANYONE CAN BEAR TO EAT A SLICE OF CAKE OR A CHOCOLATE

BACK TO NATURE

You only need to look at **MILLIONS** of years of evolution to see that the way we eat now goes against **NATURE**.

Watch an **ANIMAL** approach food...

...and you'll see that it's **INSTINCT**, not **INTELLECT**, that helps it choose the right source of nourishment.

We have the same **INSTINCTIVE** ability to discern what is **GOOD** or **BAD** for us. Remember your **FIRST TASTE** of these?

COFFEE   **ALCOHOL**   **CIGARETTES**

But we build immunity by developing a "**TASTE**" for these poisons. Actually, what is really happening is that you **LOSE** the taste!

YUK!

Most of us will have been too young to remember our first taste of **BAD SUGAR**, but those who are not **ADDICTED** will **RECOIL** from excessive sweetness.

BAD SUGAR attempts to **REPLICATE** the sweetness of **NATURAL** foods, such as **FRUIT**. Initially, this is the **ONLY** reason we are attracted to it.

But the **ONLY** reason we **CONTINUE** to eat these deadly foods is that we have been **BRAINWASHED** into seeing them as a **PLEASURE**.

It's not your **SENSES** you're **SATISFYING**; it's your **ADDICTION**.

When you **LISTEN** to your **INSTINCTS**, you know that:

THE FOODS THAT TASTE BEST *TO* YOU ARE THE FOODS THAT ARE BEST *FOR* YOU!

The **MOST NUTRITIOUS** foods are:

FRUITS   **VEGETABLES**   **NUTS & SEEDS**

Well, if THAT'S all I get to eat, you can FORGET IT!

Remember that this is not a **DIET**, but **A WAY OF LIVING**, which leads you back to your most **NATURAL** state and, more importantly, to **ENJOY** the process!

These are the **PRIMARY** foods and, ideally, will form the basis of every meal, but there **ARE** other **SECONDARY** foods which you can eat but which are **LESS NOURISHING**.

I'm not sure my INSTINCTS are tuned in anymore! How do I find the best foods?

Start **USING** your instincts and they'll soon let you know. **HANDLE** the **FOOD**. **LOOK** at it. **SMELL IT**.

Would you eat it **RAW**? If so, it's most likely **NATURAL**, **HEALTHY FOOD**.

The more **PACKAGED** the food, the more likely it is to contain **BAD SUGAR**.

Well, what about MEAT?

Because meat needs to be **COOKED**, it can be classed as a **SECONDARY FOOD**. A **LIMITED** amount is OK, but it shouldn't be the main feature.

However, a **VEGETARIAN** diet is **IDEAL**.

I'm sorry but limiting myself to a "HEALTHY DIET" sounds BORING. What about the VARIETY of other goods on offer?

Of the dozens of different breakfast cereals on the shelves, how many have you actually **TRIED**?

Hmm. A COUPLE, I guess.

Now take a stroll around the **FRUIT** and **VEG** section of the supermarket. See the huge **VARIETY** on offer.

Let's look at your **MAIN** meal...

Hmm, the **VEGETABLES**, I guess!

...which component do you **VARY** the most?

As for **CAKES** and **CONFECTIONERY** – you have a combination of pretty much the same three **BLAND** ingredients – **BUTTER**, **FLOUR**, and **SUGAR**. Not much lost in **VARIETY** or **INTEREST**, is there?

Now we need to become **ATTUNED** to another important **INSTINCT** – **WHEN** to eat.

Here's the **NORMAL** scenario until now:

Right, everyone, it's DINNER TIME!

But I'm not HUNGRY!

That's a BIG serving!

I just had a SNACK!

Eating when you're not genuinely **HUNGRY** is like overfilling your car with petrol.

And do you **TOP UP** your petrol the minute the gauge dips below **FULL**?

HUNGER is nature's way of telling us it's time to **REFUEL**, but if you eat at the **SLIGHTEST HINT** of hunger, you'll miss the full **ENJOYMENT** of a meal. The **HUNGRIER** you are, the **BETTER** food tastes.

Here is your own **FUEL GAUGE**.

HARDLY PERCEPTIBLE

TIME TO
EAT
NOW

HUNGER GAUGE

FUEL TANK
FULL.
STOP
EATING

E   F

It is also helpful to:

- **BEWARE** of "**TRIGGERS**" when you feel only **SLIGHTLY** hungry, such as the **SMELL** of food or someone talking about a good meal

- **PLAN** your meals and have healthy food handy. Many people reach for junk as a quick **STANDBY**

- **DISTRACT** yourself from mild hunger signals and you may find that they are "**FALSE ALARMS**"

- Tell yourself that you are **GUARANTEEING** yourself **REAL PLEASURE** by waiting for **TRUE HUNGER** to kick in

- Make sure you eat **SLOWLY** and **CHEW THOROUGHLY**, so that the body has time to register that it is **FULL**.

If we have this NATURAL FUEL GAUGE, why do we OVEREAT BAD SUGAR?

Our fuel gauge works **FINE**, but we have learned to **IGNORE** or **OVER-RIDE** the signals.

On **TOP** of this, processed foods have so few actual **NUTRIENTS**, our instincts have trouble even **RECOGNIZING** them as **FOOD** and it is only when we're **BLOATED** that it registers that we're **FULL**!

FALSE HUNGER is when the **LITTLE MONSTER** starts niggling for its next fix of **BAD SUGAR**. This can only be remedied by **REMOVING BAD SUGAR** completely!

YOUR NEW FAVOURITES

You've kept saying I can enjoy as much of my FAVOURITE FOOD as I like without gaining weight, but my FAVOURITE food is CHEESECAKE!

Looks like you still need a **REMINDER** of the way you have been **CONNED** into being a slave to sugar:

YOU HAVE BEEN SOLD ON A PRODUCT THAT DOES NOTHING BUT TRY TO REPLICATE THE SWEETNESS OF NATURAL FOODS SUCH AS FRUIT

Your **FIRST INSTINCT** is to favour **NATURAL, HEALTHY** foods with the highest **NUTRITIONAL VALUE**:

As a **NEWBORN**, this was **MOTHER'S MILK...**

When you were on **SOLIDS**, you favoured **FRUIT...**

...because you **INSTINCTIVELY** knew that it was the **BEST** source of all the **NUTRIENTS** you would need.

Other flavours, such as **MEAT**, were an **ACQUIRED** taste – as was **SUGAR**, which was added in large amounts or through **PROCESSING**, but left to your own devices, the **NATURAL** taste was for **FRUIT**.

Think of all those "sweet" foods and drinks you think are so **TASTY** – they are only "tasty" because of the **FRUIT-FLAVOURED** additives!

How would they taste if a **MEAT** flavour was added?

PTOO!

If you want further proof that **FRUIT** is the "**KING**" of foods, look no further than our nearest cousin in terms of sharing **DNA**, the **CHIMPANZEE**.

Chimps eat both plants and meat, but apart from using meat mostly to demonstrate their prowess, they tend to prefer **FRUIT** or **LEAVES**.

There are **MANY** good reasons to make **FRUIT** your first choice:

- **NO PREPARATION NEEDED**
- **NO WASTE**
- **NUTRITIOUS**
- **SATISFYING**
- **CONTAINS WATER**
- **STAYS COOL IN HOT WEATHER**
- **TASTE**
- **HUGE VARIETY**

But HEALTH FOODS are more EXPENSIVE!

And **COST DOESN'T COME INTO IT!**

ONE SNACK BAR　　　　**TWO APPLES**

SAME PRICE!

Before you were **BRAINWASHED**, fruit was your **NATURAL** choice. It is your **BODY'S NATURAL** favourite. The only reason there is **DOUBT** is because the food industry has done everything in its power to **BRAINWASH** you into thinking **OTHERWISE!**

The only **"PLEASURE"** derived from **BAD SUGAR** is the **TEMPORARY** relief of feeding the **LITTLE MONSTER** his **"FIX"** and briefly shutting up the **BIG MONSTER**. It's the **ILLUSION** that **BAD SUGAR** does something for you.

But you know **BETTER** now, don't you?

NO SUBSTITUTES

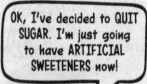

OK, I've decided to QUIT SUGAR. I'm just going to have ARTIFICIAL SWEETENERS now!

NO! This will actually force you **DEEPER** into the **TRAP!**

HOW?

Let's take a look at the **"SUBSITUTE THEORY"** by seeing how it works (or rather, **DOESN'T** work!) with other addictions such as **SMOKING.**

NRT (Nicotine Replacement Therapy – which should be called Nicotine **MAINTENANCE** Therapy!) – is based on the idea that the biggest problem with

addiction is the **PHYSICAL** cravings and if you keep them quiet, you can break the "**HABIT**" and once you've broken the habit, you can then tackle the **PHYSICAL** withdrawal by incrementally reducing your intake.

There are two fatal **FLAWS** in this approach:

1. IT'S NOT A *HABIT*; IT'S AN ADDICTION!

2. THAT *PHYSICAL WITHDRAWAL* IS SEEN AS THE BIGGEST OBSTACLE

Let's address these:

1. Smokers reach for a cigarette and go through all the rituals of smoking, not out of **HABIT** but in order to get their **FIX** of nicotine.

Substitutes **KEEP** you addicted because you're still feeding **BOTH MONSTERS** – the **LITTLE MONSTER** gets his **TOP-UP** of the drug on demand and the **BIG MONSTER** continues to convince you that the sweetness is something you **NEED**.

2. Reducing your intake bit by bit only keeps you wanting **MORE** because you feel **DEPRIVED**.

It is not really any different to using **WILLPOWER**.

Killing off the **LITTLE MONSTER** is **EASY** once you remove the **DESIRE** to smoke (the **BIG MONSTER**).

With *Easyway*, by removing the **DESIRE** first, any withdrawals are **MINOR** and gone in a matter of **DAYS**.

In fact, the **LITTLE MONSTER'S** mild death throes may have you **REJOICING** because they're an indication that your **SLAVERY** is coming to an **END**!

A LOOK AT ARTIFICIAL SWEETENERS

- They lower the hormone *leptin*, which helps regulate appetite and metabolism, so you tend to eat more and do not absorb your food effectively.

- Powdered fructose or "fruit sugar", although derived from fruit, has all the fibre and nutrients removed in processing.

- Research suggests that these substances can actually be extremely harmful.

Why would you want to make naturally sweet food sweeter than it is? Why try to dress up **SECOND-RATE**, otherwise **BLAND** food when **REAL** food is available? The only reason you need extra sweetness is **ADDICTION**.

QUIT!

IT'S THE ONLY *REAL* ALTERNATIVE

REMOVING ALL DOUBTS

But the FOOD INDUSTRY is HUGE! It's just IMPOSSIBLE to reverse the BRAINWASHING when it's EVERYWHERE I look!

We're not asking you to **CHANGE THE WORLD**! You are doing this **FOR YOURSELF**!

Put **EVERYONE** and **EVERYTHING** else aside and seek your **OWN** happiness and wellbeing through this process.

YOUR success is the only thing that **MATTERS**. It doesn't matter what **OTHER** sugar addicts are doing – **PITY** them because they're still **TRAPPED** and you've found a way out of **PRISON**!

It's surprisingly **EASY** to reverse the **BRAINWASHING**. You just need to keep an **OPEN MIND** and do the following:

DISCOVER, APPRECIATE, and **EXPLORE NATURAL FOODS.**

SEE THEM AS THE WONDERFUL, BENEFICIAL PACKAGES THEY ARE.

SMELL THEM.

FEEL YOUR MOUTH WATERING.

TRY SOME EXOTIC NEW FRUITS AND VEGETABLES.

NOTICE THE HUGE VARIETY AVAILABLE.

EXPERIMENT WITH GREAT NEW RECIPES.

- SEE YOUR OLD "FAVOURITES" FOR WHAT THEY ARE – *JUNK!*

ANALYSE THE TRUTH BEHIND THE ADS.

LOOK AT THE PACKAGING AND HOW IT CONS YOU.

Processed food tastes **BLAND** without added ingredients. See for yourself:

TRY RICE OR PASTA OR A POTATO ON ITS OWN.

EAT BREAD WITHOUT ANY TOPPING.

How do these compare to **FRESH, SUCCULENT FRUIT?**

I GET IT, but I'm still afraid that I'll feel as though I'm MISSING OUT!

Let's put these **FEARS** to **REST**.

What are they?

111

OCCASIONS just won't be the same!

If you take away the **PEOPLE** or replace them with people you **DISLIKE**, is the **FOOD** going to make any **DIFFERENCE**?

FALSE PLEASURE is the **TEMPORARY** relief of feeding the **LITTLE MONSTER** his fix, only to feel **BLOATED** and **GUILTY** later. It has nothing to do with the occasion.

The **GENUINE PLEASURE** of spending time with friends is **LASTING** and has no **DOWNSIDE**.

My chocolate bar makes the boring commute home BEARABLE!

That "treat" doesn't last the **WHOLE** journey, does it? What **THEN**? Eat **MORE**? Why not have **FRUIT** or **NUTS** instead?

The **CRAVING** for **SUGAR CAUSES** the restlessness – not the **COMMUTE**!

But so many of those foods have been **STAPLES** in our diet! How will we feel **FULL** without them?

The very pursuit of feeling "**FULL**" on foods with little or no nutritional value is not only **POINTLESS** but is a clear symptom of your **ADDICTION!**

Well, I'm reluctant to **DENY** my **FAMILY!**

BURGER FRIES

Now you know **BETTER**, do you really want to feed them **POISONOUS, ADDICTIVE JUNK** that has **NO** nutritional value?

113

Picture in your mind all the **POTATOES, BREAD, PASTA,** and **CAKES** you consume in a year. Imagine how **HARD** your body has to work to process these **UNNATURAL, UNHEALTHY,** and **ADDICTIVE** foods instead of the **NATURAL, HEALTHY** foods that are in abundance.

When you get rid of all those **CARBS** and **SUGAR** from your diet, it will feel like a **HUGE WEIGHT** has been lifted, not only from your body but also from your mind.

You have been **BRAINWASHED** into thinking that you "love" **BAD SUGAR**, but be very clear – **BAD SUGAR DOES NOT LOVE YOU!**

The so-called "pleasures" you associate with **BAD SUGAR** are no more than **BRAINWASHING** and **ADDICTION.** When you free yourself from these two **MONSTERS**, you will be amazed at how much more **ENJOYABLE** life will become. You'll feel more **RELAXED, HEALTHIER,** and more **ENERGETIC.**

Best of all, you'll feel **GOOD ABOUT YOURSELF!**

YES! You have **NOTHING** to **LOSE** and **EVERYTHING** to **GAIN**!

And **REMEMBER** – you can…

EAT AS MUCH OF YOUR FAVOURITE FOODS AS YOU LIKE, WHENEVER YOU WANT, AS OFTEN AS YOU WANT, AND BE THE EXACT WEIGHT THAT YOU WANT TO BE, WITHOUT DIETING, SPECIAL EXERCISE, USING WILLPOWER, OR FEELING DEPRIVED!

GUIDELINES

Now is a good time to do a little **REVISION** with this **SUMMARY** of the key principles we have covered:

- Your inability to control your **BAD SUGAR** intake was because of **ADDICTION** which resulted from **CONSUMPTION** and **BRAINWASHING**.

- Because this was **IMPOSED** upon you from **BIRTH**, you had no **INSIGHT** and therefore no **CHOICE**. Now you **DO**!

- Quitting **EASILY** and **PERMANENTLY** means killing off **BOTH** monsters – the **LITTLE MONSTER** and his demands for a fix and the **BIG MONSTER** who tells you there's some **PLEASURE** or **CRUTCH** in having **BAD SUGAR**.

- Killing off the **LITTLE MONSTER** is **EASY** once you've eliminated the **BIG MONSTER**. It's the **BRAINWASHING** that causes the panic.

- Forget **WILLPOWER** – you only need it if you are in **CONFLICT**. Once you realize that **BAD SUGAR** does **NOTHING** for you, there's nothing to **BATTLE** against!

- Wild animals in harmony with their **INSTINCTS** do not gain weight despite an abundance of food. Let your own **INSTINCTS** now guide you, instead of the **MIND** and its tricks.

- You do not have an **ADDICTIVE PERSONALITY**. You became **ADDICTED** to an **ADDICTIVE SUBSTANCE**.

- You didn't realize you were **BRAINWASHED**. With an **OPEN MIND**, you can **REVERSE** the process.

- You don't need **SUBSTITUTES**. They make it **HARDER** to **QUIT** and keep you **HOOKED**.

- You have **NOTHING TO FEAR!** You will sacrifice **NOTHING** except being a **SLAVE!**

- You will feel **FITTER, HEALTHIER,** and **BETTER** than ever without **BAD SUGAR**.

If there are any points you're still **STUCK** on, go back, and look at them again. It's all here – you just need to "GET IT" to be **FREE**.

Now let's look at the key principles of eating **IN TUNE** with **NATURE'S GUIDE**:

- Your new **FAVOURITES** are **FRUIT, VEGETABLES, NUTS,** and **SEEDS**. They are **EASY TO DIGEST,** have all the **NUTRIENTS** you need, and taste **GREAT!**

- The best way to **SATISFY** your hunger is to eat **RAW, HIGH-WATER-CONTENT FOOD**.

- **AVOID** excessive **SALT**, such as heavily salted nuts and seeds.

- **AVOID** all **PROCESSED FOODS**. Reconsider any food you wouldn't eat **RAW** or without **DRESSINGS**.

- **LIMIT** meat and fish as these are less easily digested. **REVERSE** your plate – make meat the **SIDE DISH** instead of the main ingredient.

- **ELIMINATE** all **REFINED SUGARS** and **STARCHY CARBS**, including **RICE, PASTA, BREAD,** and **POTATOES**. All these cause a spike in **BLOOD SUGAR**.

- **ENERGY DRINKS** are **TWO POISONS** in one – **BAD SUGAR** combined with **CAFFEINE!** No one – especially **CHILDREN** – should want that! This is simply **ADVERTISING** at its **MOST PERNICIOUS**.

- **EAT** when you're **GENUINELY HUNGRY** and eating will be truly **PLEASURABLE**. **OVEREATING** happens when you're not really **HUNGRY**.

- Eat **SLOWLY**, so that your body has time to **REGISTER** the nutrients and **STOP** when you're **FULL**.

- Forget aiming for an **IDEAL WEIGHT**. You'll know when you feel **GOOD**.

- **DIETING** doesn't work because you feel **DEPRIVED** and have to exercise **WILLPOWER**.

Juicing or blending fruit alone can create a **SUGAR BOMB**!

Juicing also tends to eliminate the **FIBRE** you need.

And avoid **PRE-PACKAGED** juices
– they're **FULL** of sugar!

When you eat fruit in its **NATURAL** state, the sugar is **BALANCED** by the fibre, so your liver can easily **METABOLIZE** it.

> What about those STATE-OF-THE-ART blenders? Don't they RETAIN the fibre?

Unfortunately, this method tends to **SHRED** the fibre, therefore rendering it **INEFFECTIVE**.

Also **BEWARE** of **DRIED FRUIT** as the dehydration process **CONCENTRATES** the sugar. There's nothing natural about it. And you never eat just **ONE**, do you?

Smoothies of **VEGETABLES** with only small amounts of **FRUIT** are better options, but the whole **FRUIT** or **VEGETABLE** is the most **NUTRITIOUS** and **NATURAL** choice.

> I'm sorry, but I don't RELISH the thought of eating just FRUIT and VEGETABLES!

There is another helpful **GUIDE** to eating, which allows some "secondary" foods, while your natural eating instincts are restored: the **GL** and **GI INDEX**.

GI is the **GLYCAEMIC INDEX** which shows the amount by which foods raise **BLOOD SUGAR**.

GL is the **GLYCAEMIC *LOAD* INDEX**, which indicates the proportion of **CARBOHYDRATE** in foods. This is a more accurate guide to rises in **BLOOD SUGAR**.

For example, watermelon has a **HIGH GI** of **72** but a **LOW GL** of **4**!

Foods with a low **GL** are **BEST**.
LOW GL is 0–10; MEDIUM is 11–19; HIGH is 20 + (you can easily find the GI and GL indexes online).

MEDIUM to **HIGH GL** foods are to be **AVOIDED**.

MEDIUM GI   HIGH GI

**INSTRUCTION: GL IS ONLY A GUIDE. TO
BE SURE YOU'RE FREE OF BAD SUGAR
ADDICTION, CUT OUT REFINED SUGAR ITEMS
AND PROCESSED OR STARCHY CARB ITEMS
(which includes pretty much all ready-made
and processed foods)**

Can I have **CHEESE** and **DAIRY PRODUCTS**?

You can have **DAIRY PRODUCTS** in **MODERATION**, but if you want to dramatically change your **BODY SHAPE** and **WEIGHT**, you'll need to **LIMIT** your intake.

A lot of **BAD SUGAR** products rely on **DAIRY** to make them **PALATABLE**, so if you're not eating **BREAD**, **PASTA**, or **CEREAL**, you won't be eating as much **DAIRY**.

Do I have to give up **ALCOHOL** ?

Firstly, you'll only think you're "giving up" something if you feel you're being **DEPRIVED**!

Alcohol can affect **BLOOD SUGAR** – especially **HEAVY** or **BINGE** drinking or drinking on an **EMPTY STOMACH**, so it's best to **AVOID** it, but an **OCCASIONAL** drink won't interfere too much.

If you have **PROBLEMS** going without **ALCOHOL**, you might want to read _Allen Carr's Easy Way to Control Alcohol_, _Allen Carr's Stop Drinking Now_, or attend one of our centres.

What if I FORGET or just go a little CRAZY one day and eat some BAD SUGAR? Does that mean I'll be HOOKED again?

Unlike other addictions, your body can process an **OCCASIONAL** sugar lapse without you getting **HOOKED** again.

If you have effectively dealt with the **BIG MONSTER** of **ADDICTION**, you'll know you gained no **PLEASURE** from it and happily resume your **SUGAR-FREE** life!

If I'm waiting until I'm REALLY hungry, WHEN should I eat? THREE MEALS a DAY fits my ROUTINE!

Well, the **GOOD NEWS** is that this fits with the **NATURAL DIGESTIVE CYCLE** too – as long as you don't **GRAZE** in between!

Get **ENOUGH NUTRIENTS** from the **RIGHT FOODS** and you won't have the desire to **SNACK.**

Basically, hunger is fairly **FLEXIBLE** and if you're **BUSY**, you can eat when it's **CONVENIENT**, but the **LONGER** you wait to eat, the more **PLEASURABLE** eating will be!

123

WITHDRAWAL

There are two things you should know about **WITHDRAWAL:**

1. **YOU HAVE ALREADY EXPERIENCED BAD SUGAR WITHDRAWALS** *EVERY DAY OF YOUR LIFE*!

2. **IT'S ONLY GRUELLING IF YOU THINK YOU'RE BEING** *DEPRIVED*!

When you eat **BAD SUGAR**, there is a **SPIKE** in **BLOOD SUGAR** then a **LOW**.

This is when the **LITTLE MONSTER** starts crying out for **MORE** and this is **ALL THERE IS** to that "terrible thing" called **WITHDRAWAL**!

Think about it – these sensations are actually **SUBTLE** and barely **PERCEPTIBLE**, aren't they?

It's the **BIG MONSTER** – the **PSYCHOLOGICAL** aspect of addiction that makes you think you're being **DEPRIVED** – that causes any **SUFFERING**.

It's not the **PHYSICAL** addiction that causes **SEVERE WITHDRAWAL** – it's the **PSYCHOLOGICAL** addiction! Let's see this in **ACTION**:

If you have adequately dealt with the **BIG MONSTER** by realizing that **BAD SUGAR** gives you **NO PLEASURE, BENEFIT**, or **ENJOYMENT**, at worst you'll experience maybe a little restlessness, fatigue, or a mild headache – hardly **UNBEARABLE!**

In fact, even if you have a few mild symptoms, aren't they worth living with for a **FEW DAYS** in return for your **FREEDOM**?

BAD SUGAR *CAUSES* THESE FEELINGS – IT DOESN'T *RELIEVE* THEM AND IF YOU CONTINUE TO CONSUME BAD SUGAR, YOU'LL HAVE WITHDRAWALS FOR THE *REST OF YOUR LIFE*!

Actually, you can **ENJOY** this mild **FEELING OF WITHDRAWAL** as part of the process of **FREEING YOURSELF**!

Huh? Whatever do you MEAN?

The **LITTLE MONSTER** was created the first time you consumed **BAD SUGAR** and was kept **ALIVE** each time you consumed **MORE**. He **COULD** have kept **TAUNTING** you for **YEARS** if you'd kept eating **BAD SUGAR**, but now he's about to **STARVE** and afterwards he won't **EVER** bother you **AGAIN**! Isn't that **GREAT**?

Oh, he'll try to **TEMPT** you in his **DEATH THROES**...

FEED ME...

Enjoy these **DEATH THROES**. The **TYRANT** who has run your life for so long is **DYING**!

When can I say I'm **CURED**?

If you have used the **_Easyway_** method, from the very moment you finish your last **BAD SUGAR** meal, you're **FREE**!

If you had tried **WILLPOWER** instead, you might have reached a point where you realized you hadn't thought about **BAD SUGAR** for a while and assumed that you were **CURED**. Your first temptation might have been to **CELEBRATE** – with **BAD SUGAR**! And so the cycle goes on and on. The **BIG MONSTER** is still very much **ALIVE**!

You, however, will be saying:

YIPPEE, I'M **FREE**!!

Soon, you will be having your **LAST BAD SUGAR** meal! Don't **WAIT**! Why stay in **PRISON** any longer than you have to?

You're about to **GAIN** so much:

- **EATING WILL BE A GENUINE PLEASURE**
- **NO MORE MOOD SWINGS**
- **YOUR BODY SHAPE WILL IMPROVE**
- **MORE ENERGY, CONFIDENCE, and MONEY**
- **YOUR SELF-RESPECT WILL BE RESTORED**
- **YOU'LL FEEL STRONGER, MENTALLY AND PHYSICALLY**

Isn't that something to be **EXCITED** about?

IMPORTANT ADVICE FOR THOSE WITH TYPE 2 DIABETES CONTROLLED BY MEDICATION OR THOSE WHO TAKE BLOOD PRESSURE MEDICATION OR ANY OTHER MEDICATION THAT COULD BE AFFECTED BY DIET AND/OR DRAMATIC WEIGHT LOSS

If you are on medication, talk to your doctor. This is particularly important because they should be involved in monitoring and adjusting your medicines.

However, you may need to be firm with your doctor and they may be resistant to your plan to cure your condition by adjusting your diet.

More and more doctors are discovering that not only is this possible but it's also incredibly easy.

Explain to your doctor what you intend to do and recruit their support.

Ask them to recommend how you can monitor and adjust your medication in line with your plan.

IF NOT NOW, WHEN?

Unless you have already quit **BAD SUGAR**, the **FINAL BAD SUGAR MEAL** is an important ritual to mark your **FREEDOM** from **ADDICTION**.

Those trying to quit using **WILLPOWER** will often choose a day to quit based on a **KNEE-JERK** reaction to suffering the effects of **OVERINDULGENCE** such as:

A HEALTH SCARE   **POST-HOLIDAY BLOWOUTS**

Of course, once the **MOTIVATION** wears off, the **DESIRE** for **SUGAR** returns, and so does the **BATTLE**. People can spend forever in discomfort, waiting for **TIME** to cure the problem, while the cravings remain.

Luckily for you, having removed the **DESIRE** in the first place, you can choose **ANY DAY** to free yourself. Why not make it **TODAY**?

THERE IS NO BETTER TIME TO STOP THAN RIGHT NOW!

That's **NORMAL** – it's the same for anyone about to take a **BIG STEP** in their lives. Think of yourself as an athlete preparing for a big race you **WANT** to run and are really **DETERMINED** to **WIN**!

You'll be **AMAZED** at how wonderful you'll feel when you finally accept that you don't need **BAD SUGAR** in your life and that dark cloud of **ADDICTION** is lifted from your shoulders!

If you have followed and understood everything in this book so far, you'll have come to the **OBVIOUS CONCLUSION –**

THERE IS NO NEED TO CONSUME BAD SUGAR!

I'm not entirely sure I'm "THERE"!

Do you still think **BAD SUGAR** does **ANYTHING** for you?

Do you still think that living without **BAD SUGAR** will leave you **DEPRIVED?**

Look at it **THIS** way – the only alternative is **NEVER BEING FREE!**

No, but the thought of NEVER having it again... I don't know...

STOP THINKING: AND START THINKING:

If you are still having **DOUBTS** – please go back and re-read the relevant chapters, keeping an **OPEN MIND!**

GREAT! But you need to be **100 PER CENT** certain – so let's remove any **LINGERING DOUBTS**, once and for all!

- You're not making a **SACRIFICE** – you're gaining **FREEDOM!**

- An occasional **SPLURGE** on **BAD SUGAR** means you're still **HOOKED** on the idea that there's some **PLEASURE** in it! You must realize there's **NOT!**

- You are not a **BAD SUGAR ADDICT** by nature – you were **CONNED** into consuming it.

- You need not be influenced by others who are still **ADDICTED. THEY** are losing out, not **YOU.**

- Ignore **BAD ADVICE** that recommends **PROCESSED** or **STARCHY CARBS** as part of a healthy diet – it's not **TRUE** that they're healthy.

It's now time for the **LAST BAD SUGAR MEAL** you will **EVER** eat.

(If you've already **STOPPED** eating **BAD SUGAR** – well done! You can skip straight to the **VOW** on the next page.)

Prepare your old "favourites" – whether a **SNACK** or a full **MEAL.**

How do they **SEEM** now?

As you **EAT**, remind yourself of all you now know about **BAD SUGAR**:

- IT IS HEAVY, BLAND, AND UNNATURAL
- IT GIVES YOU NO PLEASURE OR CRUTCH
- IT DOESN'T RELIEVE STRESS OR ANXIETY – IT CAUSES IT
- THE ONLY REASON YOU THOUGHT YOU NEEDED IT WAS BECAUSE YOU ATE IT IN THE FIRST PLACE

Now make a **VOW**.

This ritual is **IMPORTANT**. Eat what you prepared... all of it! The memory of it will always stay with you in the future and remind you of how **TRAPPED** you were by **BAD SUGAR**.

CONGRATULATIONS! YOU'VE WON! SAVOUR YOUR VICTORY!

YOU'RE SUGAR-FREE

YOU'VE DONE IT! How do you **FEEL?**

You may be confused or concerned that on a **BAD** day your thoughts return to **BAD SUGAR**. Don't **WORRY** – it's normal. Recognize that this is a remnant of the days when you turned to **BAD SUGAR** at the slightest **HINT** of stress and that you don't need to do that any more!

NEVER QUESTION YOUR DECISION TO FREE YOURSELF FROM THE TYRANNY OF BAD SUGAR ADDICTION! DO NOT PUT YOURSELF THROUGH THAT MISERY AGAIN!

One of the great **BENEFITS** of curing your addiction is that there's so much more to **ENJOY** that you previously missed because the need to feed your addiction completely **TOOK OVER.**

You can now embrace **REAL PLEASURES** such as:

- **SPENDING TIME WITH LOVED ONES**
- **ENJOYING YOUR WORK MORE WITHOUT THE DISTRACTION**
- **ADMIRING YOUR INNER STRENGTH**

Do you have any **MEAL** suggestions?

Certainly – nothing can beat beautiful, fresh **FRUIT** for breakfast, but **EGGS** are fine as a **SECONDARY FOOD** with some **SALAD**.

LUNCH or **DINNER** can be healthy **SALADS**, but make sure you enjoy exploring the huge **VARIETY** of **VEGETABLES** and **FRUITS** in different **COMBINATIONS**.

Become **INTERESTED** in your food. Enjoy choosing the **JUICIEST, FRESHEST** foods available.

A small handful of **NUTS** or **SEEDS** is a lovely end to a meal.

You will soon find that this way of eating becomes **SECOND NATURE**.

I think I'd also like to have **COOKED FOOD** sometimes though!

Having the **OCCASIONAL** fry-up or steak meal is fine, but remember to make the **VEGETABLES** and salad the "hero" on the plate, with meat as a **MINOR** component.

You don't always have to eat vegetables raw. It's just a **HANDY GUIDE** – "Would I eat that raw?" Mind you, the more raw food you eat, the better.

What about EXERCISE?

Exercise is great if that's your inclination, but it's important to make it a **PLEASURE**, not a **CHORE!**

Once you are operating within your **NATURAL CYCLE** and there is a balance between **INPUT** and **OUTPUT** of your **FUEL**, you will stay **FIT** and **HEALTHY** without any **SPECIAL EFFORT**.

So, there is nothing more for you to do except to go out and live the life you were meant to – one of **FREEDOM** and **WELLBEING!**

And remember:

YOU'RE WORTH ONLY THE BEST!

THE INSTRUCTIONS

1. **KEEP AN OPEN MIND**

2. **READ THE ENTIRE BOOK**

3. **DON'T SKIP ANY STEPS**

4. **START OFF WITH A FEELING OF EXCITEMENT AND ELATION**

5. **DISREGARD ANY PRECONCEIVED WEIGHT TARGET!**

6. **AVOID EATING WHEN YOU'RE NOT HUNGRY**

7. **IGNORE THE ADVICE OF ANYONE WHO CLAIMS TO HAVE QUIT BY WILLPOWER!**

8. **CUT OUT REFINED SUGAR ITEMS, AS WELL AS PROCESSED AND STARCHY CARB ITEMS**

YIPPEE,
I'M FREE!

TELL ALLEN CARR'S EASYWAY
ORGANISATION THAT YOU'VE ESCAPED

Leave a comment on www.allencarr.com, like our
Facebook page www.facebook.com/AllenCarr, or write to
the Worldwide Head Office address shown below.

ALLEN CARR'S EASYWAY CENTRES

The list opposite indicates the countries where Allen
Carr's Easyway To Stop Smoking Centres are currently
operational. Check www.allencarr.com for latest additions
to this list.

The success rate at the centres, based on the three month
money-back guarantee, is over 90 per cent.

Selected centres also offer sessions that deal with alcohol,
other drugs, and weight issues. Please check with your
nearest centre for details.

Allen Carr's Easyway guarantees that you will find it easy
to stop at the centres or your money back.

ALLEN CARR'S EASYWAY

Worldwide Head Office
Park House, 14 Pepys Road, Raynes Park,
London SW20 8NH ENGLAND
Tel: +44 (0)20 8944 7761
Email: mail@allencarr.com
Website: www.allencarr.com

Worldwide Press Office

Contact: John Dicey
Tel: +44 (0)7970 88 44 52
Email: media@allencarr.com

AUSTRALIA	NEW ZEALAND
AUSTRIA	NORWAY
BELGIUM	PERU
BRAZIL	POLAND
BULGARIA	PORTUGAL
CANADA	REPUBLIC OF IRELAND
CHILE	ROMANIA
CYPRUS	RUSSIA
DENMARK	SAUDI ARABIA
ESTONIA	SERBIA
FINLAND	SINGAPORE
FRANCE	SLOVENIA
GERMANY	SOUTH AFRICA
GREECE	SOUTH KOREA
GUATEMALA	SPAIN
HONG KONG	SWEDEN
HUNGARY	SWITZERLAND
INDIA	TURKEY
IRAN	U.A.E.
ISRAEL	UNITED KINGDOM
ITALY	USA
JAPAN	
LEBANON	
MAURITIUS	
MEXICO	
NETHERLANDS	

Visit www.allencarr.com to access your nearest centre's contact details.

THE ILLUSTRATED Easyway

OTHER ALLEN CARR PUBLICATIONS

Allen Carr's revolutionary Easyway method is available in a wide
variety of formats, including digitally as audiobooks and ebooks, and
has been successfully applied to a broad range of subjects.
For more information about Easyway publications, please visit
shop.allencarr.com

Good Sugar Bad Sugar

The Easy Way to Quit
Sugar

The Easy Way to Quit
Emotional Eating

Lose Weight Now

The Easy Way for Women
to Lose Weight

No More Diets

The Easy Way to Lose
Weight

Stop Smoking Now

Quit Smoking Boot Camp

Your Personal
Stop Smoking Plan

The Illustrated Easy Way
to Stop Smoking

Finally Free!

Allen Carr's Easy Way for
Women to Quit Smoking

The Illustrated Easy
Way for Women to Stop
Smoking

Smoking Sucks (Parent
Guide with 16 page pull-
out comic)

The Little Book of Quitting
Smoking

How to Be a Happy
Nonsmoker

No More Ashtrays